In this story you will learn about the **short u** vowel sound. Can you find these words and sound them out?

sun   duck   fun   run
up   jumped   pups

Here are some review sight words:

the   was   in   are   too

Here are some new sight words:

like

Here are some fun words:

puddle   quack   woof

The sun was up.

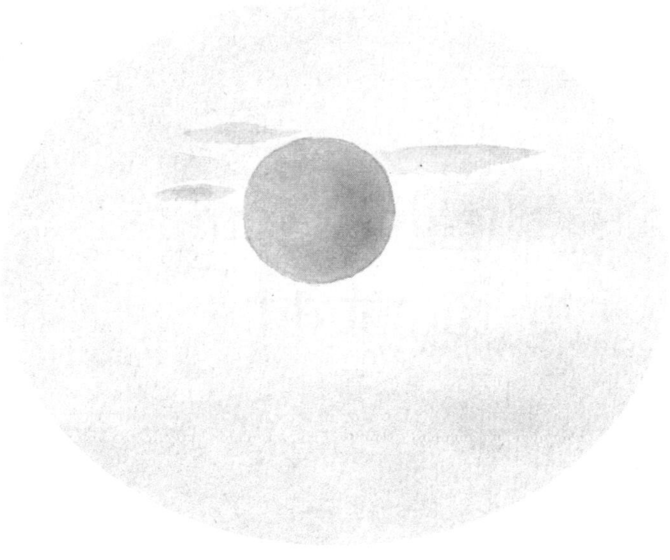

The duck jumped
in the puddle.

Quack!

Puddles are fun.

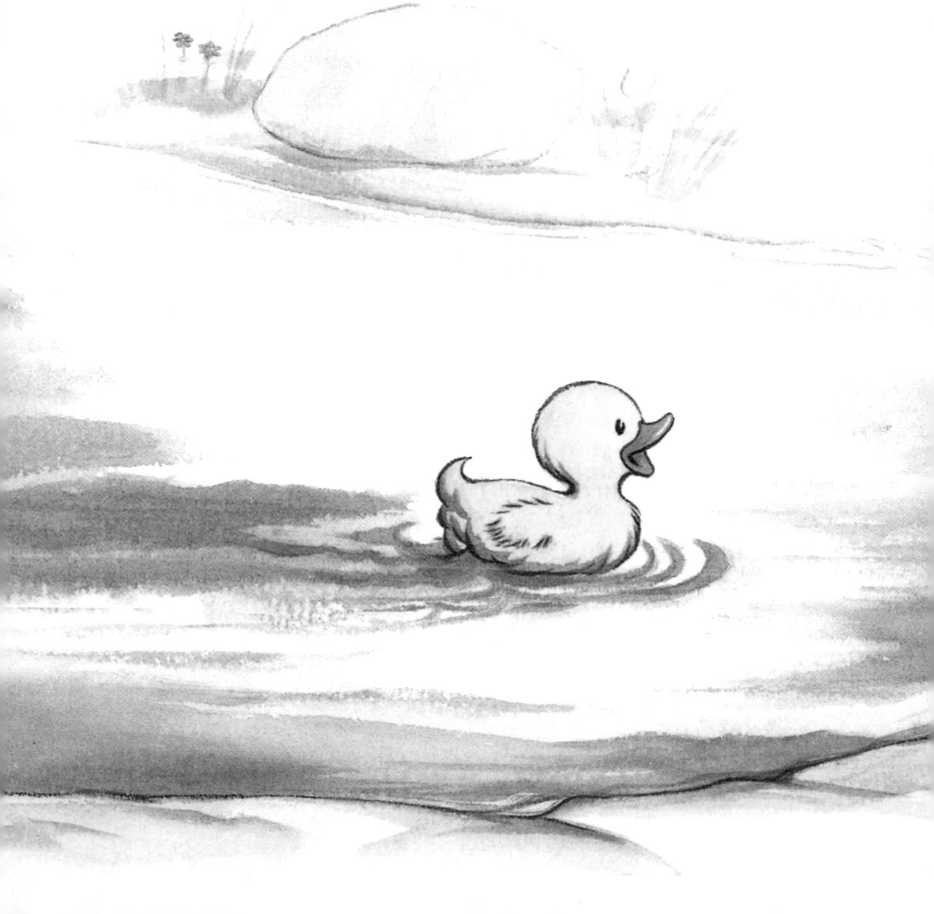

Woof!

Pups like puddles, too.

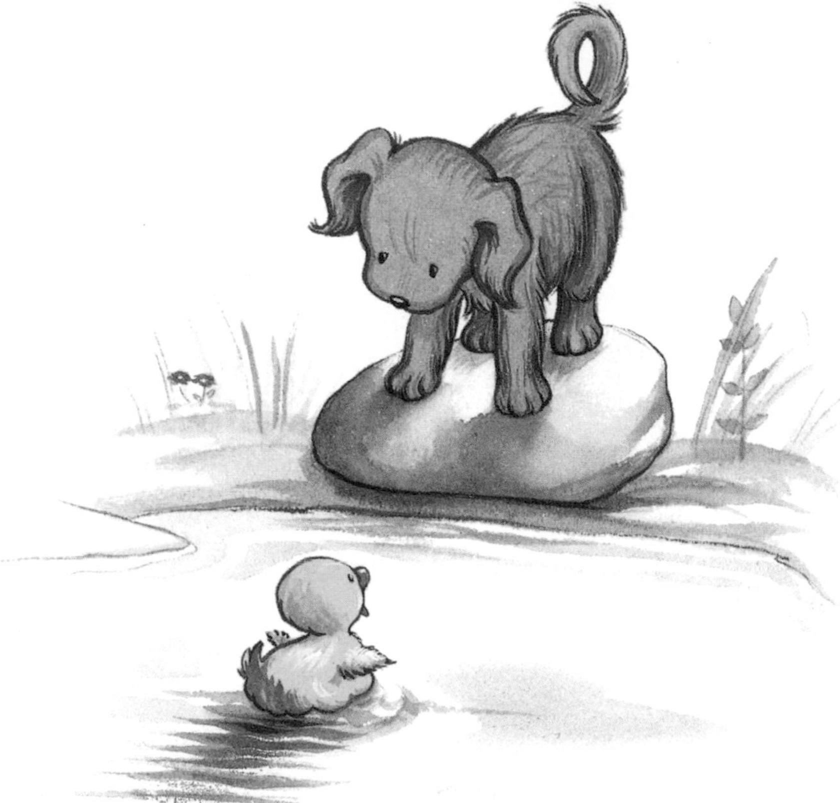

Run, run, run, jump!

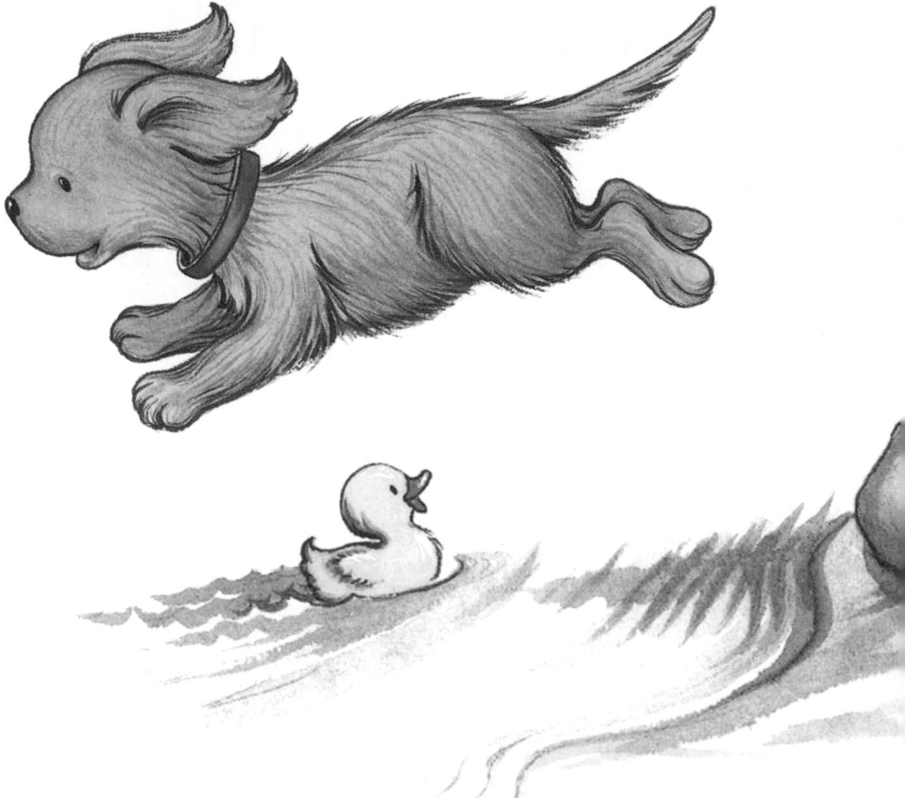

Duck fun. Pup fun.

Puddle fun!

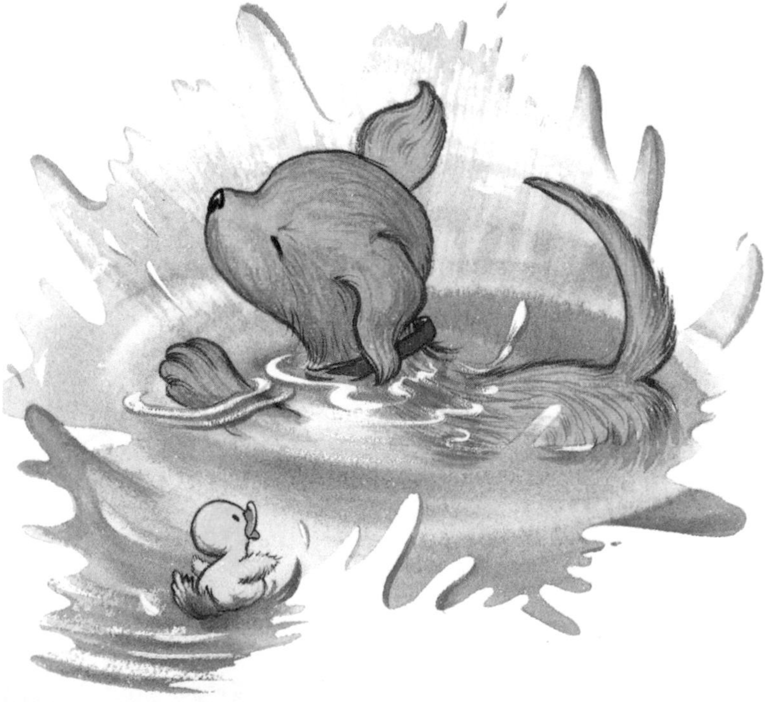